My Favorite COLOR

PURPLE

A Crabtree Roots Book

AMY CULLIFORD

CRABTREE
Publishing Company
www.crabtreebooks.com

School-to-Home Support for Caregivers and Teachers

This book helps children grow by letting them practice reading. Here are a few guiding questions to help the reader with building his or her comprehension skills. Possible answers appear here in red.

Before Reading:
- What do I think this book is about?
  - *This book is about the color purple.*
  - *This book is about things that are purple.*

- What do I want to learn about this topic?
  - *I want to learn what foods are purple.*
  - *I want to learn about shades of purple.*

During Reading:
- I wonder why...
  - *I wonder why plums are purple.*
  - *I wonder why some flowers are purple.*
- What have I learned so far?
  - *I have learned that grapes are purple.*
  - *I have learned that many flowers are purple.*

After Reading:
- What details did I learn about this topic?
  - *I have learned that there are many shades of purple.*
  - *I have learned that plums are purple.*

- Read the book again and look for the vocabulary words.
  - *I see the word **grapes** on page 7 and the word **plum** on page 10. The other vocabulary words are found on page 14.*

I see purple.

I see a purple **flower**.

5

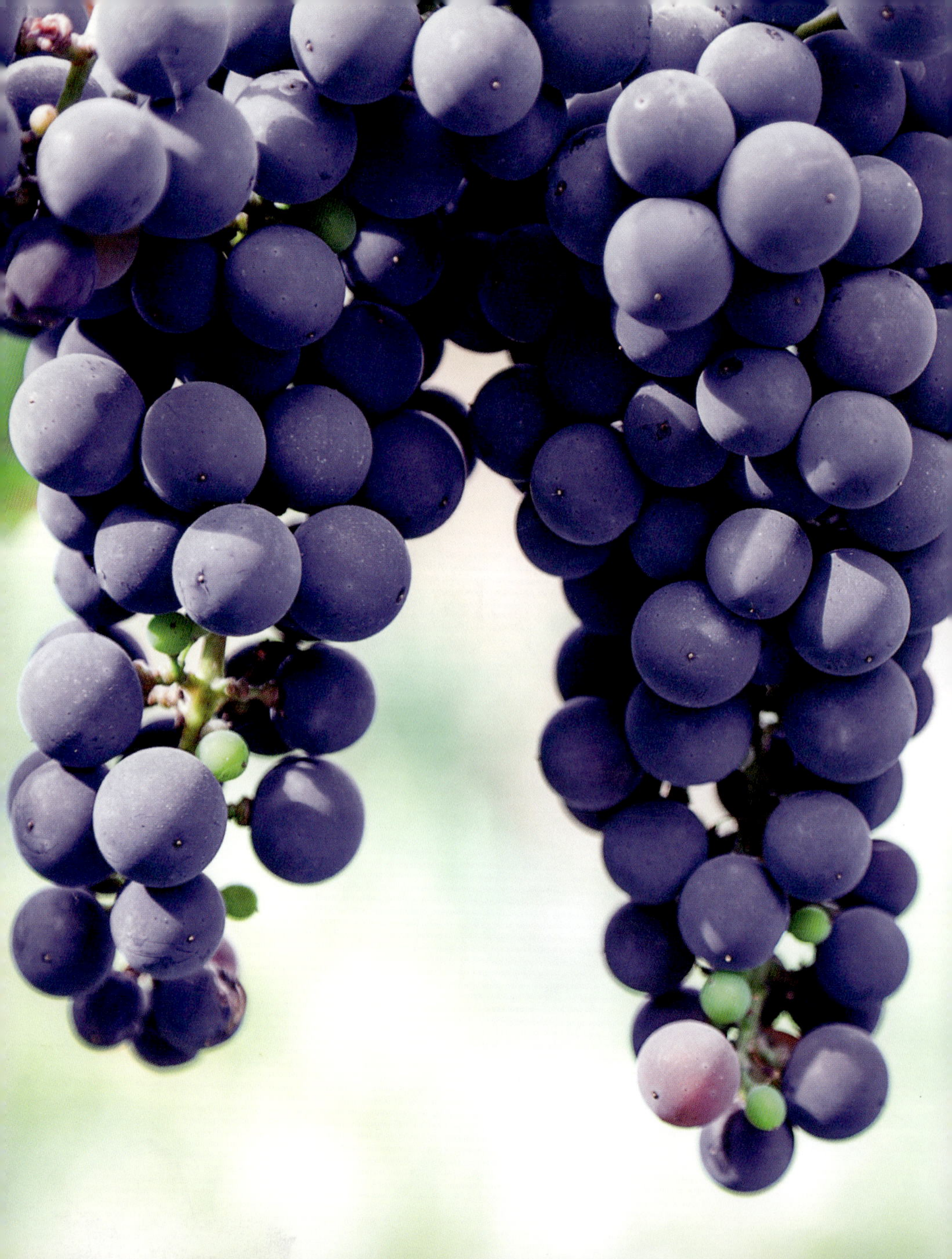

I see purple **grapes**.

I see a purple **jacket**.

9

I see a purple **plum**.

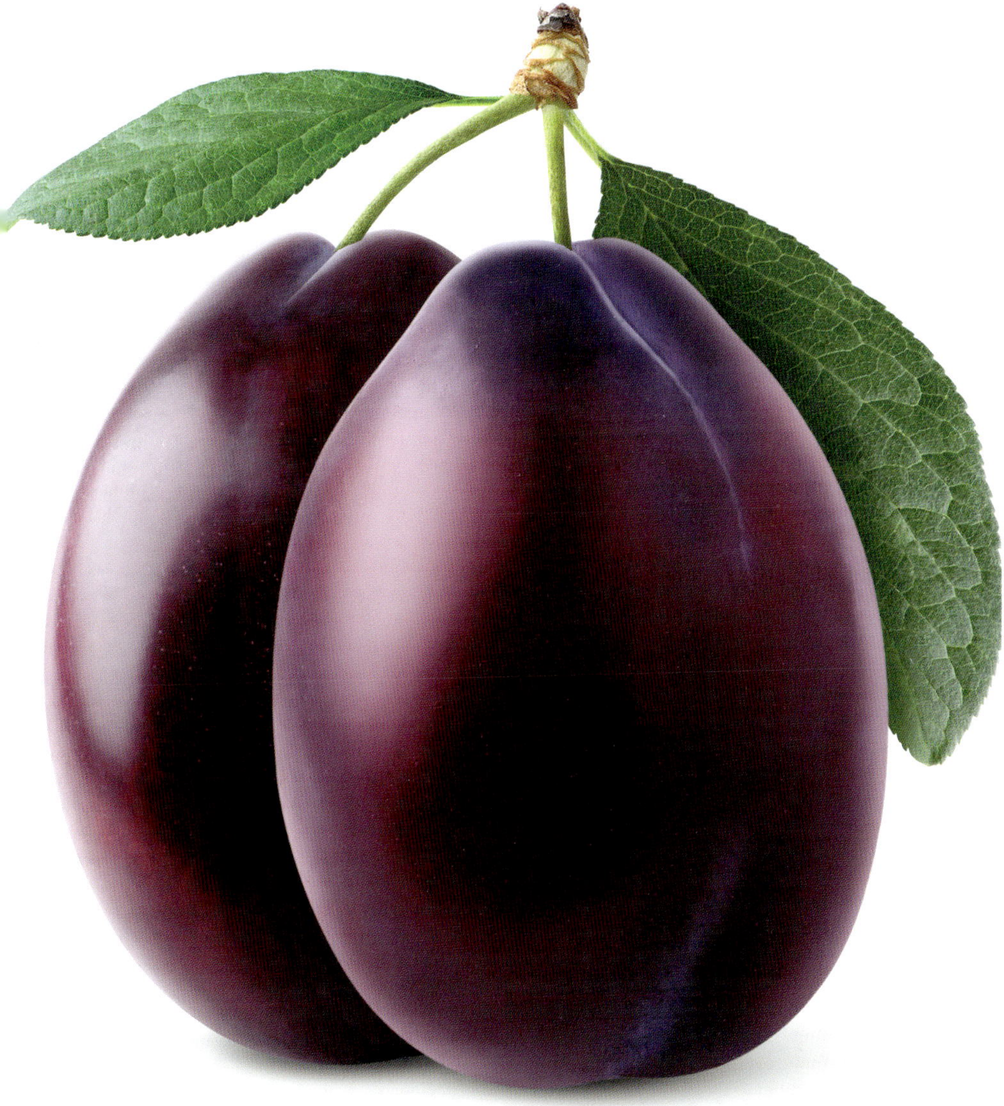

What do you see that is purple?

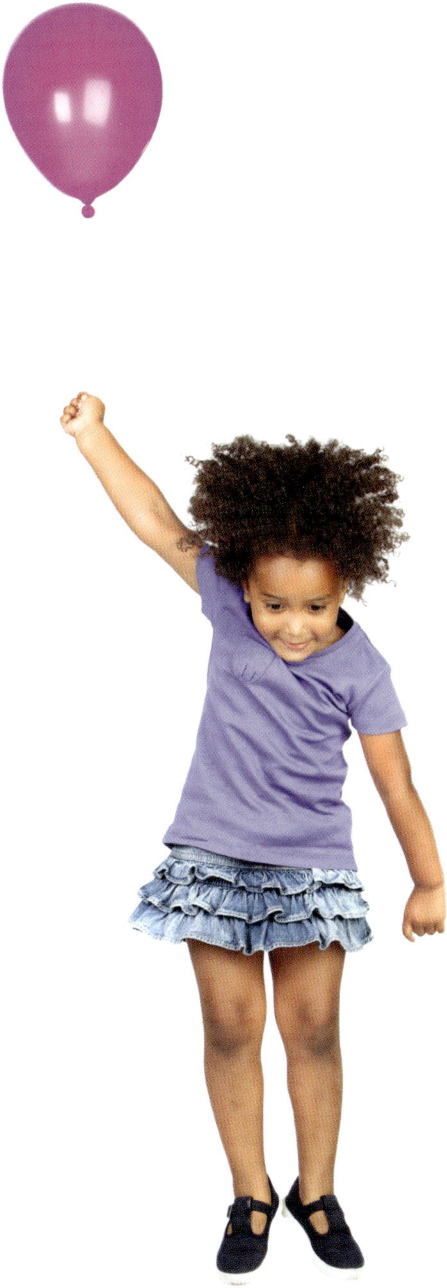

Word List

Sight Words

a	is	what
do	purple	you
I	see	

Words to Know

flower

grapes

jacket

plum

29 Words

I see purple.

I see a purple **flower**.

I see purple **grapes**.

I see a purple **jacket**.

I see a purple **plum**.

What do you see that is purple?

My Favorite COLOR PURPLE

Written by: Amy Culliford
Designed by: Rhea Wallace
Series Development: James Earley
Proofreader: Kathy Middleton
Educational Consultant:
　Christina Lemke M.Ed.

Photographs:
Shutterstock: Tarasyuk Igor: cover; freya-photographer: p. 1; CobraCZ: p. 3; Vilor: p. 5, 14; Dmitrij Skorobogatov: p. 6, 14; Artem Avetisyan: p. 9, 14; Anna Kucherova: p. 11, 14; Gelphi: p. 13

Library and Archives Canada Cataloguing in Publication
Title: Purple / Amy Culliford.
Names: Culliford, Amy, 1992- author.
Description: Series statement: My favorite color | "A Crabtree roots book".
Identifiers: Canadiana (print) 20200384015 | Canadiana (ebook) 20200384023 | ISBN 9781427134691 (hardcover) | ISBN 9781427132604 (softcover) | ISBN 9781427132666 (HTML)
Subjects: LCSH: Purple—Juvenile literature.
Classification: LCC QC495.5 .C854 2021 | DDC j535.6—dc23

Library of Congress Cataloging-in-Publication Data
Title: Purple / Amy Culliford.
Description: New York, NY : Crabtree Publishing Company, [2021] | Series: My favorite color ; a Crabtree roots book | Includes index.
Identifiers: LCCN 2020050172 (print) | LCCN 2020050173 (ebook) | ISBN 9781427134691 (hardcover) | ISBN 9781427132604 (paperback) | ISBN 9781427132666 (ebook)
Subjects: LCSH: Purple--Juvenile literature. | Colors--Juvenile literature.
Classification: LCC QC495.5 .C856 2021 (print) | LCC QC495.5 (ebook) | DDC 535.6--dc23
LC record available at https://lccn.loc.gov/2020050172
LC ebook record available at https://lccn.loc.gov/2020050173

Crabtree Publishing Company
www.crabtreebooks.com　　1-800-387-7650

Printed in the U.S.A./022021/CG20201130

Copyright © 2021 **CRABTREE PUBLISHING COMPANY**

All rights reserved. No part of this publication may be reproduced, stored in a retrieval system or be transmitted in any form or by any means, electronic, mechanical, photocopying, recording, or otherwise, without the prior written permission of Crabtree Publishing Company. In Canada: We acknowledge the financial support of the Government of Canada through the Canada Book Fund for our publishing activities.

Published in the United States
Crabtree Publishing
347 Fifth Avenue, Suite 1402-145
New York, NY, 10016

Published in Canada
Crabtree Publishing
616 Welland Ave.
St. Catharines, Ontario L2M 5V6